Andrii Titov
Ondřej Životský

Amorphous bilayered CoFeCrSiB ribbons: microstructure and magnetism

Andrii Titov
Ondřej Životský

Amorphous bilayered CoFeCrSiB ribbons: microstructure and magnetism

LAP LAMBERT Academic Publishing

Impressum / Imprint
Bibliografische Information der Deutschen Nationalbibliothek: Die Deutsche Nationalbibliothek verzeichnet diese Publikation in der Deutschen Nationalbibliografie; detaillierte bibliografische Daten sind im Internet über http://dnb.d-nb.de abrufbar.
Alle in diesem Buch genannten Marken und Produktnamen unterliegen warenzeichen-, marken- oder patentrechtlichem Schutz bzw. sind Warenzeichen oder eingetragene Warenzeichen der jeweiligen Inhaber. Die Wiedergabe von Marken, Produktnamen, Gebrauchsnamen, Handelsnamen, Warenbezeichnungen u.s.w. in diesem Werk berechtigt auch ohne besondere Kennzeichnung nicht zu der Annahme, dass solche Namen im Sinne der Warenzeichen- und Markenschutzgesetzgebung als frei zu betrachten wären und daher von jedermann benutzt werden dürften.

Bibliographic information published by the Deutsche Nationalbibliothek: The Deutsche Nationalbibliothek lists this publication in the Deutsche Nationalbibliografie; detailed bibliographic data are available in the Internet at http://dnb.d-nb.de.
Any brand names and product names mentioned in this book are subject to trademark, brand or patent protection and are trademarks or registered trademarks of their respective holders. The use of brand names, product names, common names, trade names, product descriptions etc. even without a particular marking in this work is in no way to be construed to mean that such names may be regarded as unrestricted in respect of trademark and brand protection legislation and could thus be used by anyone.

Coverbild / Cover image: www.ingimage.com

Verlag / Publisher:
LAP LAMBERT Academic Publishing
ist ein Imprint der / is a trademark of
OmniScriptum GmbH & Co. KG
Heinrich-Böcking-Str. 6-8, 66121 Saarbrücken, Deutschland / Germany
Email: info@lap-publishing.com

Herstellung: siehe letzte Seite /
Printed at: see last page
ISBN: 978-3-659-79873-3

VŠB   Technickal University of Ostrava

Institute of Physics & Nanotechnology Centre

Amorphous bilayered $Co_{69}Fe_2Cr_7Si_8B_{14}/Co_{59}Fe_{12}Cr_7Si_8B_{14}$ ribbons: microstructure, magnetic properties and their stability

Bc. Andrii Titov

doc. Ing. Ondřej Životský, Ph.D.

Ing. Yvona Jirásková, Ph.D.

1

Gratitude

In this place I would like to express my gratitudes to doc. Ing. Ondřej Životský, Ph.D., an adviser of my diploma work. Thanks to his professional and tactful leadership this work and attendant publications could appear.

Also, I would like to express my thanks to Ing. Yvonna Jirásková, Ph.D. and her colleagues at the Institute of Physics of Materials, AS CR Brno, for possibility to provide part of my measurements and for discussions about obtained results.

I also thank to Dr. Dušan Janičkovič and prof. Peter Švec from Slovak Academy of Sciences for preparation of the single-layered and bilayered ribbons.

This work was partially supported by the projects of the Student Grant Competition (SP 2014/27, SP 2015/168) and by the Czech Sciences Foundation (P108/11/1350).

Keywords: Bilayered ribbons, microstructure, magnetism, annealing, surface and bulk properties.

Contents

List of signs and symbols

Symbol	Means
B_{obs}	observed profile width
B_{std}	standard integral profile width
B_{struct}	structural broadening
d_{mean}	mean crystallite size
H	magnetic field intensity
H_c	coercive field
K	crystalline shape factor
M_L	longitudinal magnetization component
M_r	remanent magnetization
M_s	saturation magnetization
M_T	transversal magnetization component
R_a	arithmetical mean deviation of the surface roughness
R_q	root-mean-square deviation of the surface roughness
θ	angle of incidence of X-ray radiation
λ	wavelength of light
λ_s	magnetostriction coefficient
ϵ	lattice strain

List of abbreviations

Abbreviation	Means
AFM	Atomic Force Microscopy
AQ	As-Quenched
BL	Bi-layered
Co/Co	$Co_{69}Fe_2Cr_7Si_8B_{14}/Co_{59}Fe_{12}Cr_7Si_8B_{14}$
Co/Fe	$Co_{72.5}Si_{12.5}B_{15}/Fe_{77.5}Si_{7.5}B_{15}$
EDX	Energy Dispersive X-ray spectroscopy
FWHM	Full Width at Half Maximum
MO	Magneto-Optical
MOKE	Magneto-Optical Kerr Effect
MOKM	Magneto-Optical Kerr Microscopy
PFC	Planar Flow Casting
RT	Room Temperature
SEM	Scanning Electron Microscopy
SL	Single-Layered
TMC	Thermo-Magnetic Curves
VSM	Vibration Sample Magnetometer
XRD	X-Ray Diffraction

1 Introduction

It is well known that Fe- and Co-based amorphous and/or nanocrystalline ribbons are widely used as ultra-soft magnetic materials in various applications, among others also in sensorics [1]. About twenty years ago a few scientific groups introduced a new type of bilayered sensors based on a junction of a low magnetostrictive amorphous ribbon to a plastic sheet or to a non-magnetic carrier ribbon having a markedly higher thickness and the magnetostriction coefficient [2–4]. A stand-by connection of different materials induces either tensile or compressive stresses into the magnetic part of a sensor. As it was found the relative permeability of an amorphous ribbon depends linearly on stress. This is valid in a specific range of mechanical straining; moreover this dependence can be easily controlled by the thickness ratio of connected both sheet/ribbon and ribbon/ribbon. Such bilayered sensors with tunable sensitivity found their applications mainly in medicine, e.g. for the monitoring a lung ventilation, knee joint bending, heart activity, or chest wall displacements [2, 3]. The principal problems of this type of sensors were connected with a very poor mutual adhesion of both materials. Therefore the efforts of various laboratories were concentrated on a fabrication of monolithic amorphous bilayered ribbons. One of the possibilities has been the conventional method of planar flow casting (PFC). The efforts to use this relatively low-cost technology for production of various components were connected with attempts to combine different elements and to prepare amorphous and/or nanocrystalline materials with often unexpected and unique physical properties. The first experiments have used separate crucibles placed bumper-to bumper with melts of different compositions. Bilayered ribbons produced in such a way had strongly inhomogeneous properties of both layers. Later two crucibles were replaced by a single crucible with a partition panel dividing the crucible into two parts for two different melts. Nozzles at the bottom of crucible are close to each other and allow ejection of both melts on the rotating wheel almost simultaneously. This technique was tested for the first time in 1991 with two different precursors, FeNiB and CoFeCrSiB [5–7]. Since that time the method was significantly improved and currently it allows one to prepare bilayered ribbons of various material combinations, e.g. CoSiB/FeSiB [8], FeCuNbSiB/FeNbSiB [9] etc. A practical usage of these materials can be documented also by the FeMnSi/SiFe ribbons that were newly developed for materials convenient for ferromagnetic shape-memories [10].

The aim of the present diploma work is the investigation of the bilayered (BL) $Co_{69}Fe_2Cr_7Si_8B_{14}/Co_{59}Fe_{12}Cr_7Si_8B_{14}$ (Co/Co) ribbons prepared by modified PFC technology. The whole work consists of four chapters; Chapter 1 is devoted to an introduction into the topic, Chapter 2 provides an utmost complex analysis of the microstructure and physical properties of the as-quenched (AQ) Co/Co samples and compares them with those obtained at $Co_{72.5}Si_{12.5}B_{15}/Fe_{77.5}Si_{7.5}B_{15}$ (Co/Fe) ribbons investigated previously in my bachelor work [11]. Within the surface and bulk studies we have newly included the properties obtained along the cross-section of both bilayered ribbons. Especially method for visualization of magnetic domains at cross-section is highly innovative and enables observation of domain patterns in individual layers of bilayered ribbon from the surface towards the interface. Detected differences between both BL ribbons are evoked mainly by dissimilar magnetostrictions done by a composition of individual layers contributing to origin of strains and thereby to smaller or higher coiling of the ribbon.

Chapter 3 is devoted to the changes in the physical and microstructural properties of BL Co/Co ribbon and single-layered (SL) ribbons of corresponding compositions owing to annealing. Temperature of 423 K was chosen pursuant to the thermomagnetic curve (TMC) measurements as suitable for thermal relaxation analysis. Behaviour of magnetic and microstructural properties in dependence on the annealing time was checked by experimental methods sensitive to the bulk and surfaces. Improvement of the soft magnetic properties and only low changes in the magnetization saturation values indicates positive influence of long-time annealing.

2 As-quenched $Co_{69}Fe_2Cr_7Si_8B_{14}/Co_{59}Fe_{12}Cr_7Si_8B_{14}$ ribbons

The studied $Co_{69}Fe_2Cr_7Si_8B_{14}/Co_{59}Fe_{12}Cr_7Si_8B_{14}$ (Co/Co) as-quenched (AQ) ribbons, 36 μm thick and 8 mm wide, were prepared by the modified Planar Flow Casting (PFC) technology [9,11]. As was mentioned, the recently upgraded PFC uses one crucible divided into two parts ended by nozzles close to each other. This allows ejaculating two melts of different and/or the same compositions practically simultaneously on the rotating copper wheel and ensures better homogeneity of the layers and an interface. The $Co_{69}Fe_2Cr_7Si_8B_{14}$ side was during preparation in contact with a rotating wheel (matt or wheel side), while the opposite $Co_{59}Fe_{12}Cr_7Si_8B_{14}$ layer (shiny or air side) was in contact with the surrounding atmosphere. Results obtained on the bilayered (BL) Co/Co AQ ribbons were compared (i) with the corresponding single-layered (SL) Co-based ribbons and (ii) with our previous works [8,11] provided on the BL ribbons $Co_{72.5}Si_{12.5}B_{15}/Fe_{77.5}Si_{7.5}B_{15}$ (Co/Fe) prepared by the same technology.

The coiling of the produced BL ribbons either along their length (longitudinal direction) or width (transverse direction) is strongly influenced by the magnetostriction coefficients λ_s of the material used. Values of λ_s of individual layers measured by direct method [12] are summarized in Table 2. The bigger difference in the λ_s between CoSiB and FeSiB layers shown in Table 2 causes clearly visible coiling of the Co/Fe ribbon, whereas the Co-side forms the concave surface and the opposite Fe-side the convex surface. On the other hand only very small coiling is visible at Co/Co ribbon due to the similar composition of both layers. The coiling cannot

9

be completely neglected because of the PFC technology alone. It is to note that both BL ribbons are more brittle as compared to their SL analogs, partially due to dissimilar λ_s and partially due to a strain originating in a stand-by connection of two materials.

Table 1: Investigated bilayered ribbons and their magnetostriction coefficients.

ribbon	layer	side	λ_s (ppm)
Co/Co	$Co_{69}Fe_2Cr_7Si_8B_{14}$	wheel	-1.0
	$Co_{59}Fe_{12}Cr_7Si_8B_{14}$	air	+4.0
Co/Fe	$Co_{72.5}Si_{12.5}B_{15}$	wheel	-2.6
	$Fe_{77.5}Si_{7.5}B_{15}$	air	+32.0

In the following parts the microstructure and magnetic properties of mentioned ribbons are investigated from the viewpoint of the surface, bulk, and interface.

2.1 Microstructural properties

The microstructural properties were investigated by Scanning Electron Microscopy (SEM) and X-Ray Diffraction (XRD). A TESCAN LYRA 3XMU FEG/SEM scanning electron microscope equipped with an Oxford Instruments Energy Dispersive X-ray Analyzer X-Max 80 (EDX) and by applying an accelerating voltage of 15 kV was used for the microstructure studies and determining the element concentration profiles across the BL ribbons. It was further completed with nanoindentation measurements. An X'PERT – PRO diffractometer with Co $K\alpha$ radiation ($\lambda = 0.17902$ nm) in a Bragg-Brentano geometry was used to confirm the amorphous structure of ribbons eventually to detect nanocrystallization. The mea-

10

surements were done in an interval of 2θ between 40° to 80° with a step of 0.008° at room temperature from both sides of BL samples and the corresponding (wheel or air) sides of SL samples. The penetration depth of γ-rays is approximately 10 μm. The PI85 pico-indenter by Hysitron was used for the micro-hardness and Young's modulus measurements using the constant load of 15 mN on cross-section of the BL sample.

2.1.1 Structure

The results of the XRD measurements of the BL and SL Co/Co samples are shown in Fig. 1. It is well seen, that no sharp peaks, which are typical for crystalline structures, are present. The broad peaks at 53.43° and 53.47° represent the amorphous structure taken from the air and wheel sides of the BL Co/Co sample, respectively. Positions of peaks in the BL sample are in good agreement with positions of peaks of the SL samples of corresponding compositions. The observed small shift of both peaks (53.46° and 53.58°) to higher angles can be ascribed to experimental error. For the sake of completeness it is to note that also the structure of Co/Fe samples was fully amorphous without any signs of crystallization.

2.1.2 Cross-section

For cross-section investigations the BL samples were fixed in a special holder (Fig. 2) providing sufficient stability and compactness during loading. The grinding and polishing were done to guarantee the surface smoothness and flatness. After that the cross-section surface was covered by carbon to guarantee a good electrical conductivity.

The cross-sectional microstructure and indentations of both BL sam-

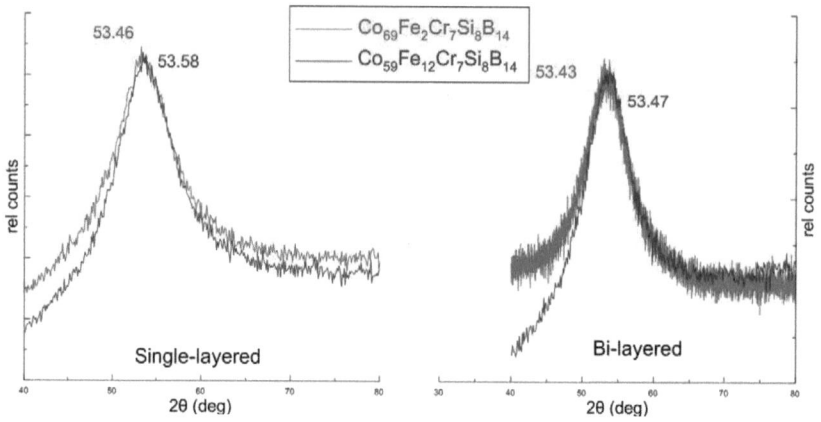

Figure 1: X–ray diffraction patterns of SL (left) and BL (right) Co/Co ribbons.

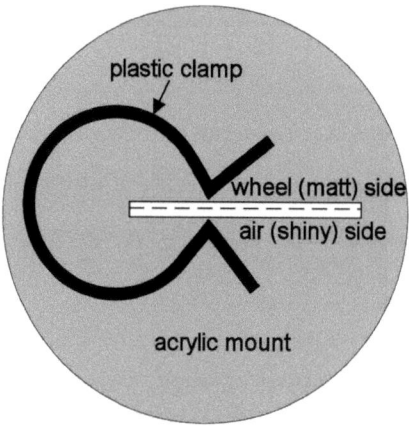

Figure 2: Schematic picture of the sample holder used for the cross-sectional structure observations and nanointendation measurements.

ples are seen in Fig. 3 on the left, while the element concentration profiles measured in steps of ≈ 2 μm and microhardness with Young's modulus are seen on the right. Co/Co sample exhibits very narrow interface almost invisible in the SEM picture (Fig. 3a) due to the similar element compositions of both layers. This reflected in a sharp change of the concentration profiles at the interface (< 1 μm) in EDX graph and by nearly step change in both hardness and Young's modulus dependence (Fig. 3 right above).

On the other hand, the interface of Co/Fe sample is much broader as seen on micrographs (Fig. 3 c, d [8]). It ranges between point 7 from the wheel side close to point 10. Nevertheless its thickness is not homogeneous along the ribbon. In some places the thickness was around 6 μm, in some only 2 μm. EDX analysis confirms much slower transition of elements across the interlayer featured by mixing of Co and Fe atoms. Both mechanical characteristics depicted in Fig. 3 (right panel, bottom) change almost continuously from one side to the other. As was shown in the previous studies [8] the origin of an interlayer of variable thickness along the ribbon volume influences mainly the bulk magnetic properties like the shape of thermomagnetic curves and/or the value of the Curie temperature. Newly, in the next subsection, we will demonstrate how the transition of the elements affects also the magnetic domain structure at the cross-section.

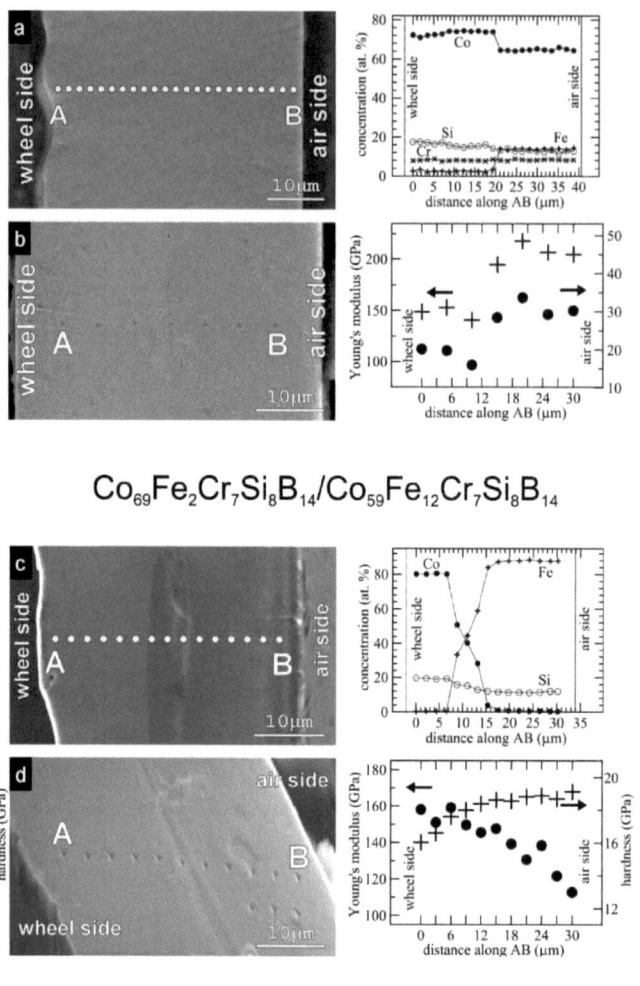

Figure 3: SEM pictures of microstructure and indentation (left), concentrations of basic elements and microhardness with Young's modulus (right) for the Co/Co (a, b) and Co/Fe (c,d) ribbons.

2.2 Magnetic properties

Surface magnetic properties of ribbons were studied using the magneto-optical Kerr effect (MOKE) experiments. We have used two MO configurations schematically depicted in Fig. 4. The first measurements (Fig. 4a) were done using s- or p-polarized red light ($\lambda = 670$ nm) that incidents the sample surface at the angle of about 45°. Measured surface hysteresis loops represent the dependence of reflected beam intensity on the applied external magnetic field. They are obtained from the surface area corresponding to the diameter of an incident laser beam (≈ 0.3 mm) and from the depth of about 20 nm. We measure the MO angle of the Kerr rotation that is proportional to the longitudinal magnetization component M_L lying in the sample plane and plane of incidence of the light. Magnetic field is generated by the air coil in the sample plane along the M_L. MOKE hysteresis loop measurements were supplemented by observation of the surface magnetic domain structure using the MO Kerr microscopy (MOKM).

For the second configuration (Fig. 4b) which enables observation of the magnetic domains at the cross-section of the BL ribbon a special sample holder was needful. The sample holder shown in the previous subsection and used for the measurements of the cross-sectional microstructure, element distribution, and mechanical characteristics was suitable also for the magnetic measurements.

The bulk magnetization curves were measured using the vibrating sample magnetometer (VSM) Microsense EV9 at maximal magnetic field ±12.5 kA/m. The thermomagnetic curves, obtained on the small samples of 3 mm in diameter, were taken by VSM manufactured by EG& G Princeton

Applied Res. Corporation. The measuring parameters: external magnetic field of 4 kA/m, temperature increase of 4 K/min, vacuum of approx. 10 mPa, and temperature range from the room temperature (RT) up to 1100 K, were maintained the same for all samples.

2.2.1 Surface

Fig. 5 shows the results of the MOKE obtained at both surfaces of the Co/Co sample which was fixed for this observation on the planar sample holder. By focusing the light on the different places on the air ribbon side almost rectangular hysteresis loops with the fast reversal and typical band magnetic domains, well visible in Fig. 5 right, were measured. This indicates that direction of the easy magnetization axis lies close to the ribbon axis. Although the sample was almost not coiled, we suppose that the mentioned way of the sample fixing was responsible for inducing the low uniaxial anisotropy. The main reason for this is the absence of fingerprint domains typically occurring at AQ ribbons that originate in the local places due to the presence of planar compressive stress. On the other hand wide band domains are the consequence of tensile stress induced by the way of sample fixing.

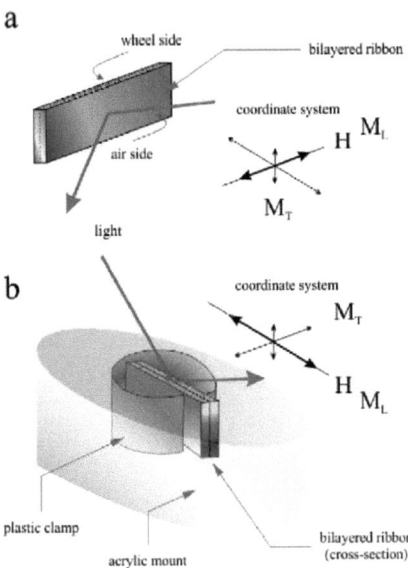

Figure 4: Magneto-optical configurations of the surface and cross-sectional studies. Subplots (a) and (b) correspond to the setting of the light focused on the surface and cross-section, respectively. M_L and M_T denote in-plane longitudinal and transversal magnetization components, parallel and perpendicular to the applied magnetic field H.

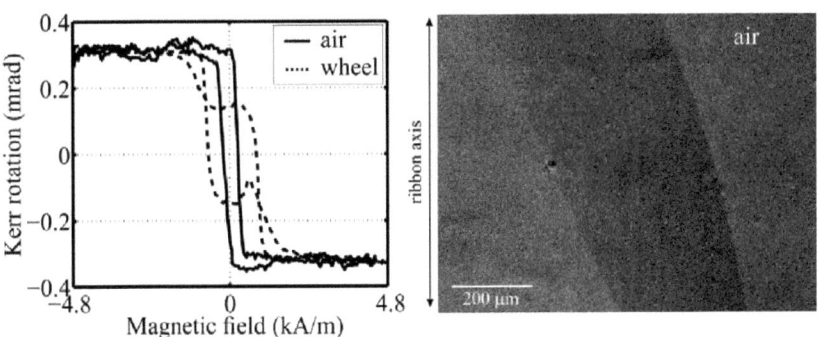

Figure 5: MOKE hysteresis loops measured at both surfaces of bilayered Co/Co ribbon (left subplot) and corresponding magnetic domains (right subplot) at air surface.

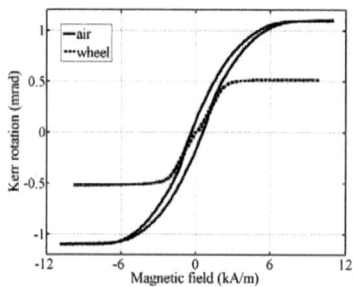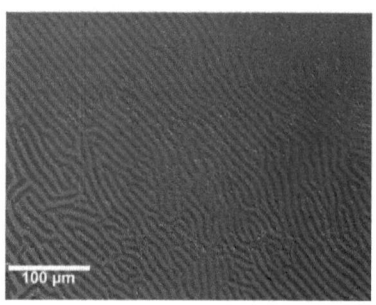

Figure 6: MOKE hysteresis loops measured at BL Co/Fe ribbon (left panel) and corresponding magnetic domain patterns observed at air surface (right panel).

More complicated hysteresis loops reflecting a presence of two different magnetic phases were observed at the wheel surface (Fig. 5 left, dotted line). The first phase is saturated approximately at the same magnetic field as the phase on the air side, while the second phase exhibits magnetically harder behaviour and the saturation at about 2.4 kA/m. A similar hysteresis loop was detected at FeSiB ribbons in the AQ state and the origin of two magnetic phases was ascribed to the presence of the FeSi and FeB amorphous clusters yielding various magnetic properties. According to this observation the formation of the (Co,Fe)Si and (Co,Fe)B amorphous clusters can be expected also at Co/Co ribbons. Magnetic domains using the MOKM were not observed at wheel side because of the markedly higher surface roughness in comparison to the opposite air side.

The surface hysteresis loops and magnetic domains of Co/Fe ribbons are shown in Fig. 6. To realize these measurements the sample has to be straightened and adherent on the planar sample holder. This has induced a compressive force and consequently the magnetic anisotropy perpendicular to the sample surface. Both hysteresis loops (taken from the wheel and air sides) clearly indicate the presence of hard magnetization axis (Fig. 6

18

left). The observed fingerprint domains (Fig. 6 right) are the consequence of the local anisotropy with easy axis perpendicular to the surface.

Magnetic anisotropy of the ribbons can be controlled by applying uniform tensile or planar compressive stress. In such a way induced anisotropy overcomes the local stresses from preparation process and ribbons are then mainly used as magnetic sensors with tunable sensitivity. Application of stress is often combined together with the ribbon annealing.

In our case the anisotropy was controlled by coiling of the SL and BL Co/Co samples into the form of toroids. The outer side of the toroid is influenced by the tensile stress in the ribbon axis, while the inner side is exposed to the compressive stress in the ribbon plane perpendicular to the ribbon axis. The strain changes linearly from the compression on the inner side of the ribbon to the extension on the outer side. However, the induced uniaxial magnetic anisotropy depends also on a sign of magnetostriction coefficient λ_s.

A situation for the SL $Co_{59}Fe_{12}Cr_7Si_8B_{14}$ sample which is characterized by positive magnetostriction can be described in the following way; if it is coiled with its air side outside, we observe an easy magnetization axis along the ribbon axis on the air surface and a hard magnetization axis along the ribbon axis on the wheel surface. This case is demonstrated in Fig. 7, where the MOKE magnetic domains from the air side are observed as a function of toroid radius. The straight ribbon (Fig. 7a) shows the presence of the wide curved and also the fingerprint domains that come from the preparation process as mentioned above. Samples were then put into the specially designed curled holder and fixed with the air side out. For the toroid radius 25 mm (Fig. 7b) the fingeprint domains completely disappeared, but the band domains are still a little bit inclined from the ribbon axis. Only when the radius of the toroid is decreased down to 13 mm (Fig. 7c), wide band domain patterns having the direction of the ribbon axis are detected. This is in agreement with our expectations.

If the sample is wound with the wheel side out, the orientations of the easy and hard axes on both surfaces shoud be mutually interchanged. Moreover, exactly opposite behaviour could be observed also for ribbons with negative magnetostriction coeficient, such as SL $Co_{69}Fe_2Cr_7Si_8B_{14}$ sample. Owing to high roughness of the wheel sides the domain observations were not possible.

An attempt to coil the BL Co/Co sample into a toroidal shape was not successfull. The reason was a brittleness of the ribbon. Nevertheless, based on our previous observations done on SL sample, it can be supposed that a winding-up the BL sample with the air side out the easy magnetization axis would have a direction along the ribbon axis on both surfaces due to dissimilar signs of magnetostriction coefficents in both layers. The stronger anisotropy should exhibit the layer with higher λ_s. It should be also mentioned that a certain uniaxial anisotropy is already induced by fixation of the sample on the planar sample holder. This anisotropy should not be changed too much by coiling into the toroidal shape. In a case of the wheel side out, it is clear that the hard magnetization axes are induced along the ribbon axis in both layers.

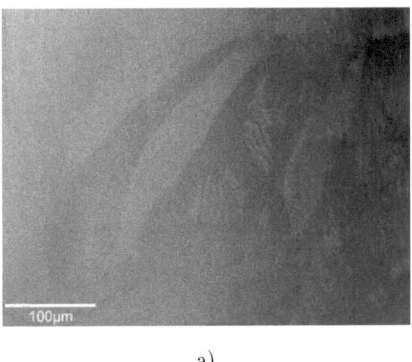

a)

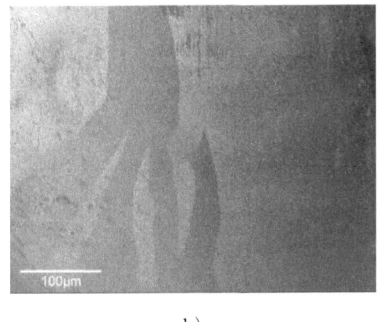

b)

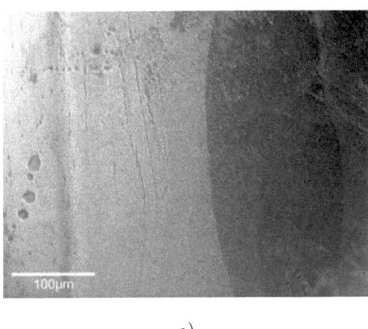

c)

Figure 7: Magnetic domain pattern observed on the air surface of straight SL $Co_{59}Fe_{12}Cr_7Si_8B_{14}$ ribbon (a) and on the air surface of the SL sample coiled into the toroid of 25 mm (b) and 13 mm (c) in radius.

2.2.2 Bulk

The bulk magnetization curves of the BL Co/Co and corresponding SL samples measured in the initial AQ amorphous state are presented in Fig. 8. The magnetic field was applied along the ribbon axis. The main magnetic parameters obtained from the hysteresis loops - remanent magnetization M_r, saturation magnetization M_s, and coercive field H_c, are summarized in Table 2.

A difference between MOKE and VSM measurements is based on dif-

21

ferent sensitivity of both methods. While the MOKE is sensitive to the surface regions accurately limited by dimensions of the laser spot and light penetration depth, the VSM detects the magnetic response averaged over the whole sample volume. The sensitivity of bulk hysteresis loop measurements to the both ribbon amorphous surfaces is therefore nearly zero. A contribution of both surfaces to the bulk magnetic properties could be higher in case of surface nanocrystallization which was not our case.

Table 2: Remanent magnetization M_r, saturation magnetization M_s, and coercive field H_c of the as-quenched BL Co/Co and both SL samples.

	BL Co/Co	SL $Co_{59}Fe_{12}Cr_7Si_8B_{14}$	SL $Co_{69}Fe_2Cr_7Si_8B_{14}$
M_r [A·m²/kg]	1.172	0.477	0.613
M_s [A·m²/kg]	47.7	57.5	35.7
H_c [A/m]	26.27	7.2	22.4

Table 3: The values of coercivity H_c, magnetic saturation M_s and remanent magnetization M_r of the Co/Fe samples.

	BL Co/Fe	SL FeSiB	SL CoSiB
M_r [A·m²/kg]	13.58	16.85	3.08
M_s [A·m²/kg]	72.2	170.3	44.4
H_c [kA/m]	3.2	3.6	0.9

From the Table 2 it is seen that value of M_s of BL ribbon falls practically in the middle between both SL ribbons. This confirms that BL sample in its initial state owns "dual" properties of both SL samples. On the other hand, the highest values of M_r and H_c correspond to the fact that low magnetic anisotropy was induced due to fixation of the ribbon on the planar sample holder.

Generally, the saturation magnetization is not structure sensitive parameter and it is determined only by chemical composition. On the other hand, the remanent magnetization and coercivity are the magnetic characteristics highly sensitive to structure and stresses relief in it. Therefore these values markedly differ for all samples. The main reasons are the presence of interlayer in the BL sample and induced magnetic anisotropy due to fixation of the samples on the planar sample holder.

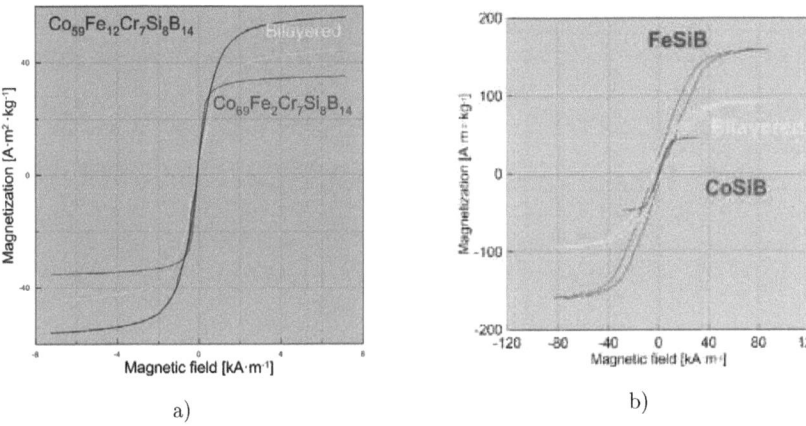

Figure 8: Bulk magnetic hysteresis loops of all three types of the Co/Co (left panel a) and Co/Fe (right panel b) samples.

The bulk thermomagnetic curves (TMC) of SL and BL Co/Co ribbons are shown in Fig. 9. Samples were heated from the room temperature up to the 1100 K and cooled back in the external magnetic field of 4 kA/m. Vacuum of about 10 mPa was used to protect the samples against oxidation. Processes of heating and cooling are depicted separately in the left and right subplot of Fig. 9.

The first decrease of magnetization on TMC of the BL sample follows the transition of the SL $Co_{69}Fe_2Cr_7Si_8B_{14}$ sample into a paramagnetic state. Observed Curie temperature of 370 K is in good agreement with that obtained in Ref. [13]. Above approximately 400 K the magnetization of the BL sample is determined by the second $Co_{59}Fe_{12}Cr_7Si_8B_{14}$ phase on the air side of the sample and it proceeds into a paramagnetic state at 475 K. The Curie temperature of the BL sample (475 K) is clearly determined by composition of the SL $Co_{59}Fe_{12}Cr_7Si_8B_{14}$ ribbon without any visible contribution of thin interface. All magnetic transitions on the TMC curves are shown in detail I in the upper middle subplot of Fig. 9.

Above 475 K all samples stay in paramagnetic state till approximately 900 K, when the magnetization of the BL and iron rich SL ribbon increases due to a starting crystallization of probably (Fe, Co)Si phase. The subplot II in Fig. 9 presents details of magnetization during the first crystallization. It is seen that shapes of both curves are similar, but the magnetization of the BL sample is markedly lower due to the presence of the second wheel layer currently in a paramagnetic state and thin interface. The second transitions of the partially crystallized BL and iron rich SL samples into the paramagnetic state occurred again at the very close temperatures: 1005 K and 1010 K, respectively.

Two magnetic transitions at the iron rich SL sample, three at BL sample, and only one at iron poor SL sample are observed on TMC at decreasing temperature. The first increase of magnetization at approximately 1000 K is connected with transition of crystalline phase in the iron rich SL and BL samples into ferromagnetic state. The next increasing steps in magnetization are connected with other crystalline phases, the composition of which are unknown at present and will be a topic of future investigations.

The changes of magnetization of the crystallized bilayered sample at decreasing temperature do not follow the thermomagnetic curves of the SL samples. All samples return back to the ferromagnetic state. An influence of crystallization on the properties of the BL sample will be a topic of next studies.

Finally we can say that thermomagnetic properties of bilayered Co/Co ribbon are mainly influenced by the air $Co_{59}Fe_{12}Cr_7Si_8B_{14}$ side. Similar conclusions were detected also in the case of BL Co/Fe sample, see Fig. 10 [8, 11], where the properties of layer prepared from iron alloy are dominated. On the other hand, much thicker interface containing CoFe-SiB phase dominated during ribbon crystallization and was responsible for bigger differences between BL and SL samples.

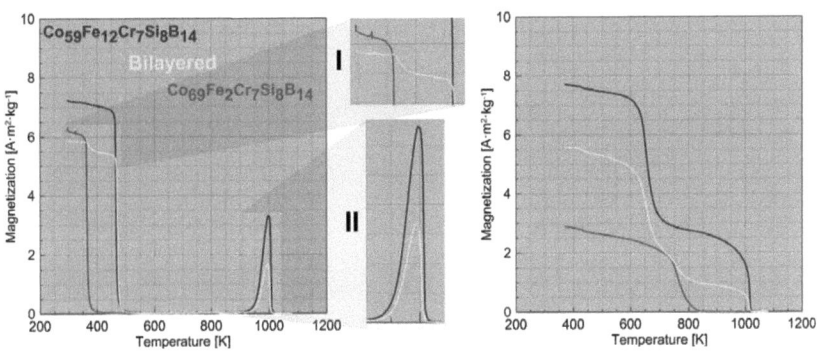

Figure 9: Thermomagnetic curves of the BL Co/Co and corresponding SL samples.

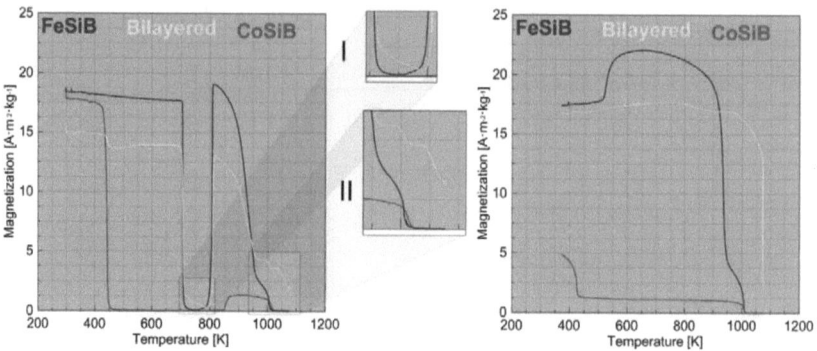

Figure 10: Thermomagmetic curves of the BL Co/Fe and corresponding SL samples.

2.2.3 Interface

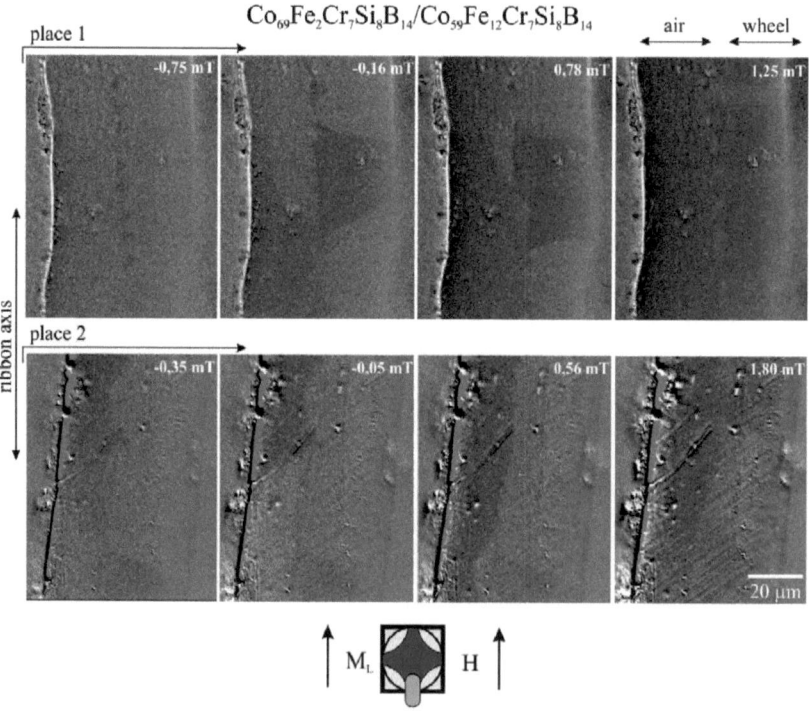

Figure 11: Domain patterns at the cross-section of the BL Co/Co sample as a function of a magnetic field applied along the ribbon axis taken from two different places [14].

Observations of magnetic domains at the cross-section of BL Co/Co and Co/Fe ribbons are shown in Figs. 11 and 12. The images in Fig. 11 present the evolution of the domains and the domain wall movements in dependence on the applied external magnetic field of different small strength taken from two chosen places. In accordance with the MOKE hysteresis loops discussed in the previous paragraphs, (i) the domains arise and change at low magnetic fields and (ii) they disappear first on the air ribbon side. Moreover, it is seen that they do not cross the interface and that they move only inside the layers. A position of the thin interface, serving as a barrier between both layers, is pointed out by a different contrast of magnetic domains. It must be stressed that domain patterns differ in every ribbon place. Moreover, their origin is positioned close to the interface (e.g. the left picture of place 1) in most of the observed places. It is also important to emphasize that the optimal and precise pictures of domains were provided by measuring the longitudinal magnetization component. The sensitivity to the transversal magnetization component has markedly decreased the magnetic contrast of domains.

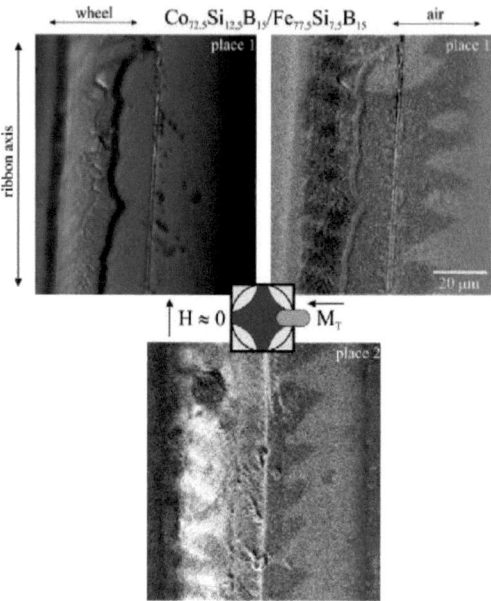

Figure 12: Magnetic domain patterns at the cross section of BL Co/Fe sample in its remnant state taken from two places. In the case of place 1, the optical microscope as well as the MOKE contrast images are presented [14].

Completely different shapes of domain patterns are visible at cross-section of Co/Fe ribbon in Fig. 12. The domains were observed in two places in remanence state. In the upper panel of Fig. 12 the images of the optical microscope and magnetic contrast taken from the same place are compared. Narrow band domains, almost perpendicular to the ribbon axis, are observed close to both surfaces. On the wheel side they remain practically unchanged when penetrating deeper below the surface, while on the air side they gradually spread towards the interface. This described behavior can be explained by different signs of λ_s and by the ribbon coiling. By fixing the sample into the sample holder, the ribbon is partially straightened. Both tensile and compressive stresses induce a transverse

28

magnetic anisotropy on both ribbon sides owing to the dissimilar signs of λ_s. Consequently, the domains are not closed along the interface as in the Co/Co ribbon but across it. Domain patterns originate also inside the thick interlayer and usually have similar contrast as those on the air side (place 1). Our assumptions are confirmed by the fact that measured domains, in agreement with previous results obtained for the Co/Fe samples, are sensitive to the transversal magnetization component M_T.

3 Annealed $Co_{69}Fe_2Cr_7Si_8B_{14}/Co_{59}Fe_{12}Cr_7Si_8B_{14}$ ribbons

This chapter is devoted to the properties of annealed BL Co/Co ribbon. As in the previous chapter the microstructure and magnetic properties are compared with properties of the corresponding SL samples. The main aim was to follow relaxation of amorphous structure and a role of an interlayer in the BL samples.

For temperature treatment the samples were cut from each ribbon. In such a way 8 sets of both SL and BL samples, approximately 8x8 mm in dimensions, were prepared. The annealing temperature of 423 K was chosen according to thermomagnetic curves measurements approximately 50 K below the magnetic transition and sufficiently below start of crystallization (Fig. 9). 8 set corresponded with 8 selected times of annealing, namely 3, 6, 12, 24, 48, 96, 192, and 384 hours. All samples were annealed in argon atmosphere to minimize surface oxidation. Each set was subsequently exposed to experimental microstructure observations and magnetic measurements.

3.1 Microstructure properties

Analogous to AQ samples, the changes in microstructure of the annealed samples were investigated by XRD at room temperature and using the same program for all samples to provide approximately the same statistics of diffraction pattern measurements. The diffractograms were taken from both sides in case of BL sample and corresponding sides at SL samples. Because they did not yield any signs of crystallization after annealing in the whole time interval, the analysis of patterns was done in such a way that the peak position of the main diffraction line and its width in the half of its maximal intensity (FWHM) was determined.

Also the surface mapping of the samples was observed by Atomic Force Microscopy (AFM) using Ntegra Prima platform (NT-MDT) at ambient conditions. The topographic image was collected from the sample area by using the semicontact/lift mode. The tip used for AFM measurements was coated with CoCr magnetic film approximately 30-40 nm thick. The curvature radius of the tip was about 40 nm.

3.1.1 Surface mapping

BL Co/Co sample was analyzed from both sides. The SL samples were analyzed from the air or wheel side corresponding to the relevant side of the BL sample. The AFM observations were done on the samples in the AQ state and after 384 hours of annealing because in this state the most visible changes could be awaited. Surface topography of all samples is presented in Figs. 13 and 14. The marked difference between AQ and annealed states is observed on the air side of both SL and BL samples. It is seen that AQ air surfaces are smooth without any visible roughness which is in good agreement with observations presented in literature [15,16]. The very small light dots could be either dust particles or very small crystalline and/or oxide nucleus present in very thin surface region and not observable by other surface sensitive experimental methods used in present work (e.g. MOKE). Nevertheless, after 384 h of annealing the surface smoothness is at first sight unchanged but some larger objects have appeared. Because it is known that the crystallization of amorphous ribbons prepared by PFC technology begins at the air surface [17], we can suppose that the long-

31

time annealing, though at low temperature, initializes the very low surface crystallization. The size of grains could be about few nm in case of the BL air side and much smaller in case of the SL air surface. The very bright dots are more likely dust particles. The main parameters of the surface roughness, arithmetical mean deviation R_a and root-mean-square deviation R_q obtained during samples mapping are summarized in the Table 4. Both parameters, R_a and R_q , have increased after annealing and confirm our speculation of initial surface crystallization contributing to a change in atomic composition on the air surfaces of both samples. In the case of wheel sides of BL and SL samples, both values R_a and R_q are approximately one or two orders higher in comparison to the air sides. Both parameters visibly decreased at the BL sample after 384 h of annealing at 423 K while at the corresponding SL sample they remained unchanged within the frame of experimental error. This could be due to relaxation of the surface structure namely annealing of free volumes and topological reordering of atoms.

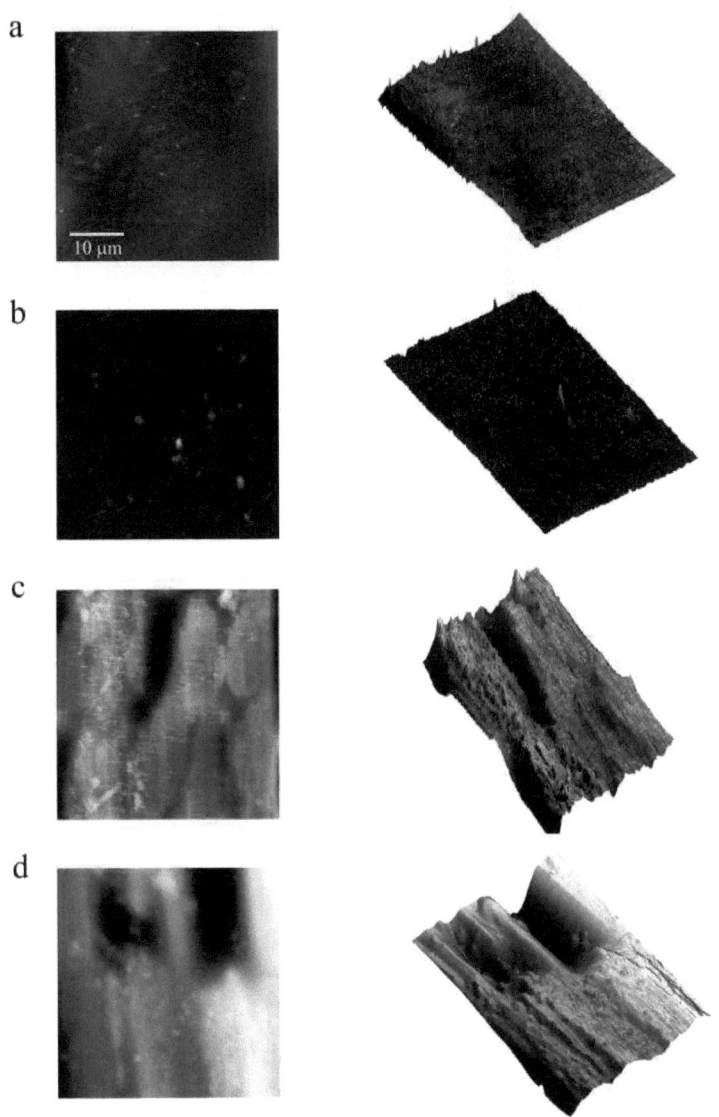

Figure 13: AFM images of the SL $Co_{59}Fe_{12}Cr_7Si_8B_{14}$ ribbon from the air surface (subplots a, b) and SL $Co_{69}Fe_2Cr_7Si_8B_{14}$ ribbon from the wheel surface (subplots c, d). Subplots a and c belong to the AQ state, while subplots b and d to the 384 h of annealing. Corresponding 3D pictures are presented in the right panel.

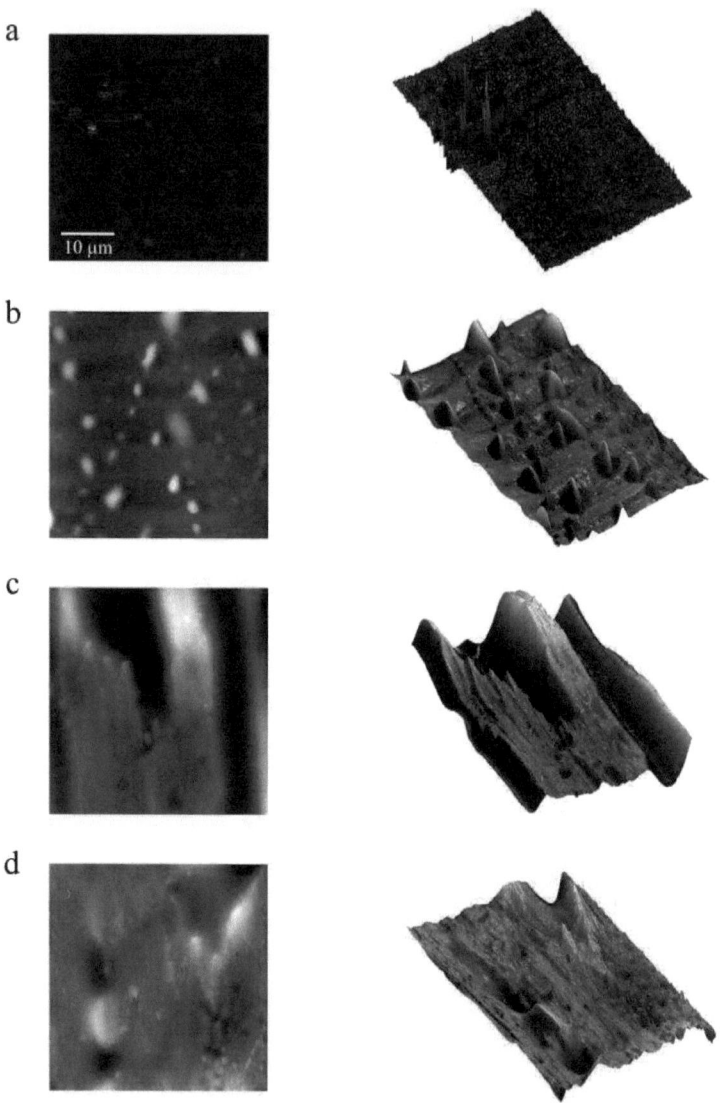

Figure 14: AFM images of the BL Co/Co ribbon from the air surface (subplots a, b) and wheel surface (subplots c, d). Subplots a and c belong to the AQ state, while subplots b and d to the 384 h of annealing. Corresponding 3D pictures are presented in the right panel.

Table 4: Parameters of the surface roughness obtained from the AFM images. R_a and R_q denote arithmetical mean and root-mean-square deviation of profile of the samples, respectively.

	BL				SL			
	air side		wheel side		$Co_{59}Fe_{12}$ air side		$Co_{69}Fe_2$ wheel side	
	AQ	annealed	AQ	annealed	AQ	annealed	AQ	annealed
R_a [nm]	2.3	18.5	241	78	2.7	15.8	102.9	103
R_q [nm]	2.9	24.9	285	96	3.5	24.9	125.5	130

3.1.2 Structure

Figs. 15 and 16 show the changes in the XRD peak position and peak width in dependence on time of annealing. Left and right panels correspond to the SL and BL samples, respectively. As it was mentioned in Chapter 1 the amorphous structure of ribbons prepared by PFC yields a certain local ordering of atoms into very small objects, clusters, which are manifested by the wide peaks in diffractograms. In certain approximation we can try to calculate the size of these clusters using Scherrer formula [18] used for calculation of the mean crystallite size in crystalline alloys:

$$d_{mean} = \frac{K\,\lambda}{B_{struct}\,cos\theta},$$

where:

- K describes the crystallite shape factor

- λ denotes wavelength used (for Co it is 0.17903 nm)

- θ defines the angle of incidence

and

- B_{struct} describes the structural broadening, which is the difference in integral profile width between a standard (std) and the sample to be analyzed (obs = FWHM - Full Width at Half Maximum): $B_{struct} = B_{obs} - B_{std}$

The K shape factor depends on type of particles and reaches the values:

Particles:	Shape factor (B values based on FWHM):
Spheres	0.89
Cubes	0.83-0.91
Tetrahedra	0.73-1.03
Octahedra	0.82-0.94

Because the form of clusters is not uniform and also not exactly known we have tried to use various values of K and B_{std} and calculate the cluster size as for AQ as for annealed samples. We have found that an influence of both parameters is in present case insignificant. The values of cluster size were very small and ranged between 1.6-1.8 nm.

The parameters, peak position and FWHM, obtained from diffractograms can be used also for determination of microstrains using tangent formula:

$$\epsilon = \frac{B_{struct}}{4 \cdot tan\ \theta},$$

where $B_{struct} = \sqrt{B_{obs}^2 - B_{std}^2}$.

The changes in microstrain with time of annealing are depicted in Fig. 17 for SL samples (left) and BL sample (right).

In Fig. 15 we can see that under annealing the peak position of the iron poor SL sample taken from the wheel side (red curve) increases and slightly oscillates with time of annealing. This could reflect that the clusters in the sample relaxed, the inter-cluster distances change, and very probably it comes also to slight chemical changes in cluster composition. At BL sample the opposite tendency is observed. On the other hand the oscillations of the peak position in the iron rich SL sample taken from the air side (blue curve) are bigger and nearly the same is observed at corresponding curve at BL sample. Based on AFM results we can speculate that the air surfaces influence the bulk properties more as a consequence of the observed surface crystallization and a higher sensitivity of the iron rich surface to oxidation. The oscillations of both parameters are visible also in Fig. 16 showing the time dependence of FWHM and in Fig. 17 where the changes of microstrains with time of annealing are depicted. Finally it can be said that obtained dependencies do not show any smooth behaviour. The reason is the complicated amorphous structure, presence of interlayer at BL sample and last but not least the sensitivity of both analyzed parameters (peak position and FWHM) to different velocity of removing the samples from the temperature zone to cool zone after individual time-steps.

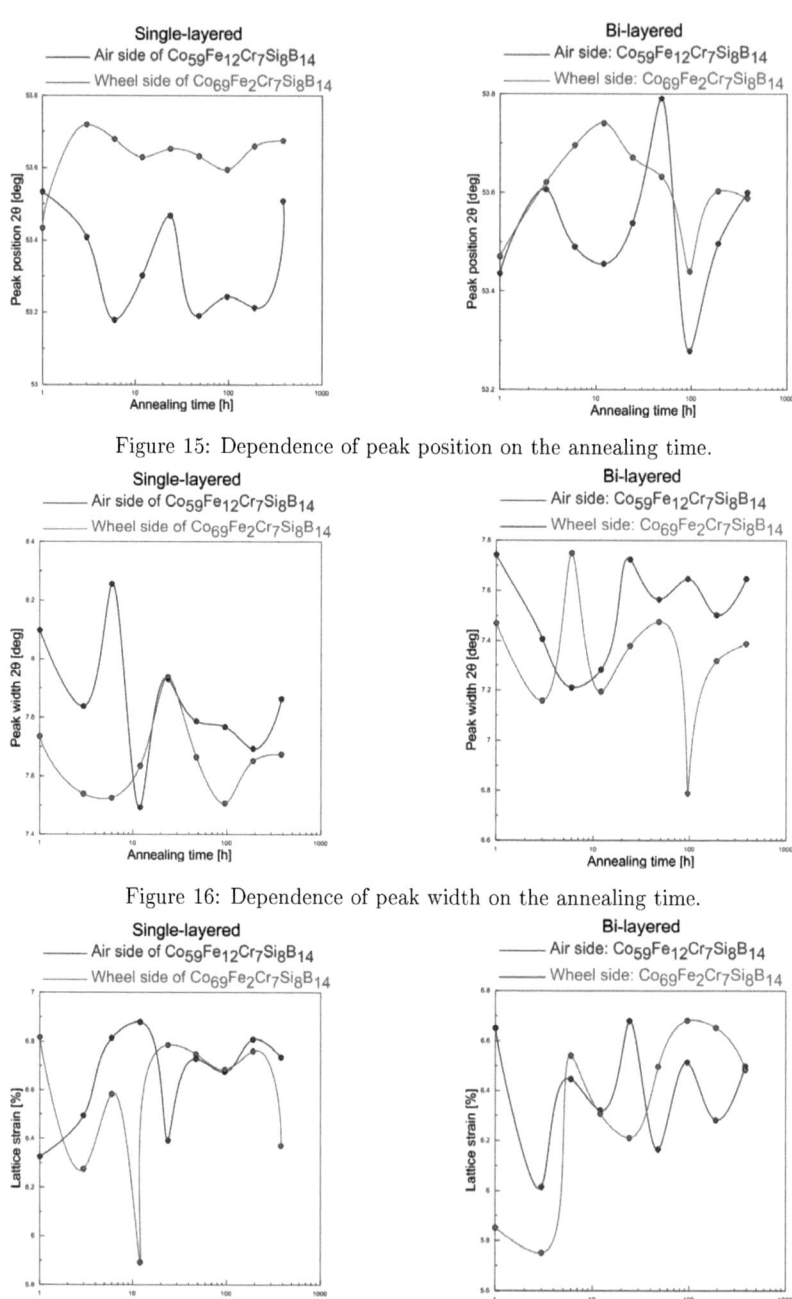

Figure 15: Dependence of peak position on the annealing time.

Figure 16: Dependence of peak width on the annealing time.

Figure 17: Dependence of lattice strain on the annealing time.

3.2 Magnetic properties

In this section the magnetic properties of annealed SL and BL Co/Co samples are compared with the AQ one presented in previous chapter. We used the same experimental techniques for surface (MOKE) and bulk (VSM) analysis, except the thermomagnetic investigations.

3.2.1 Bulk

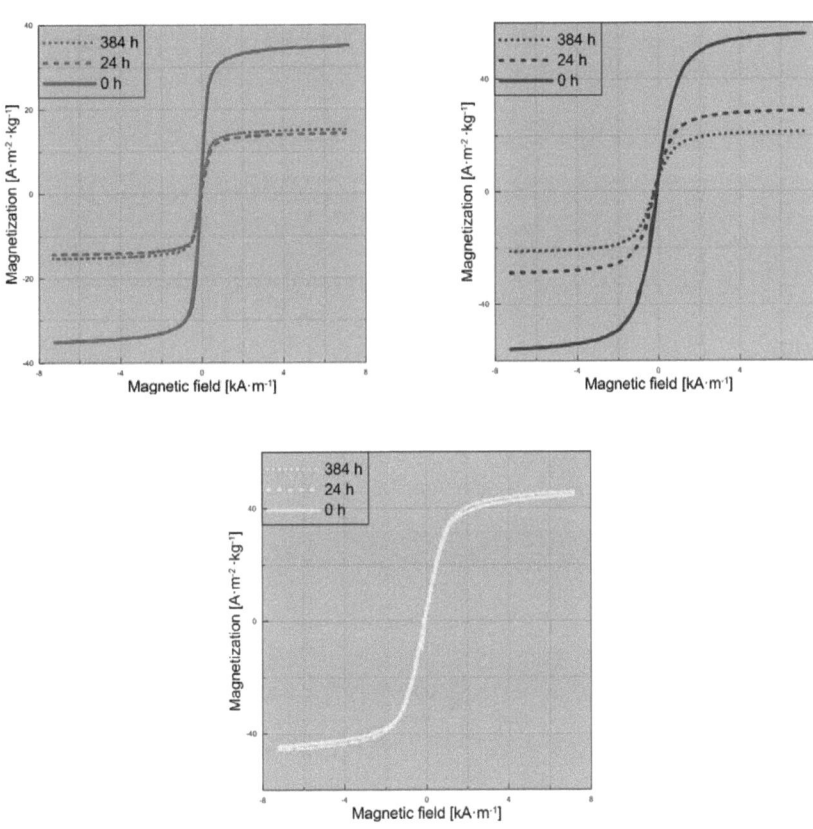

Figure 18: Magnetic bulk hysteresis curves of the SL $Co_{69}Fe_2Cr_7Si_8B_{14}$ (upper left subplot), SL $Co_{59}Fe_{12}Cr_7Si_8B_{14}$ (upper right subplot), and BL Co/Co (lower subplot) samples depicted as a function of annealing time.

Figure 18 shows the dependence of the bulk hysteresis loops on the annealing time. For better visualization we present only the curves of samples annealed at 24 h and 384 h and compare them with ribbons in AQ state (0 h). Evolution of the main magnetic parameters obtained from the curves, i.e saturation magnetization M_s, remanent magnetization M_r, and coercive field H_c, is depicted as a function of annealing time in Fig. 19.

Our results confirm that long-time annealing significantly improves bulk soft magnetic properties of BL Co/Co ribbon. This is documented by the (i) decrease of H_c from 26.3 A/m (AQ state) to 1.6 A/m (after 384 h of annealing), (ii) decrease of remanent magnetization close to zero, and (iii) practically identical value of M_s slowly fluctuating about 48 Am^2/kg during the whole annealing process. It is obvious that connection of both SL alloys, forming a thin interface, and additional annealing positively influence magnetic behavior and seems to be useful for practical applications of these materials (like sensors, memories). Both SL samples as well as BL samples exhibit decrease of M_r with increasing annealing time. A decreasing tendency is observed also at coercivity. Some fluctuations are probably connected with stresses induced owing to manipulation with samples. The saturation magnetization is influenced only at bot SL samples. It has markedly decreased during the first three hours of annealing and practically did not change up to the final annealing times. The reason could be relatively fast annealing out of free volumes present in the amorphous structure after ribbon production and homogenization the chemical composition. Contrary, the M_s of the BL sample is nearly unchanging. A decrease in M_r and H_c confirms positive influence of relaxation annealing and simultaneously documents that the low crystallization observed on the air surface of the BL and corresponding SL by AFM does not influence bulk magnetic properties in a significant way.

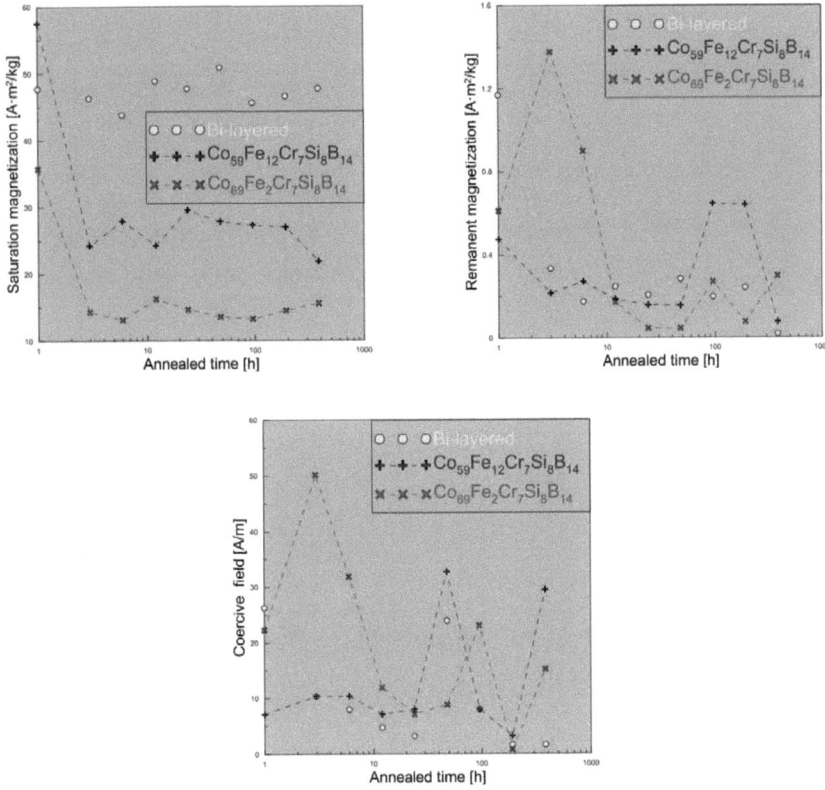

Figure 19: Evolution of the saturation magnetization (upper left subplot), remanent magnetization (upper right subplot), and coercive field (lower subplot) as a function of annealing time for both SL and BL samples.

3.2.2 Surface

Fig. 20 shows the MOKE hysteresis loops obtained at different annealing times at the air side of BL Co/Co ribbon and at corresponding sides of SL ribbons. Differences are seen at first sight. Coiling of the BL sample did not change much during temperature treatment. Therefore, most of

MOKE loops exhibits typical rectangular shape with the easy magnetization axis close to the direction of magnetic field that was applied along the ribbon axis (loops for 3 h and 384 h annealing in Fig. 20). Induced magnetic anisotropy is not caused by the annealing process, but similarly as in the case of AQ Co/Co ribbons due to fixing the sample on the planar sample holder. This process leads to the origin of wide band domains in local ribbon places as is shown in subplot a of Fig. 21. The domain wall then reflects orientation of the easy magnetization axis.

As was mentioned previously the coiling of BL sample is not identical along the ribbon. It is a reason together with local MOKE sensitivity why we can measure also different forms of hysteresis loops as it is documented at the sample after 24h of annealing. On the bottom sub-figure in Fig. 20 we can see the contributions of two magnetic phases. Central part indicates the presence of easy axis, while the hard axis is detected for applied magnetic field higher than 0.5 kA/m. Existence of both phases due to fixing is the consequence of inhomogeneous ribbon coiling in the various local places. In such a case, it is possible to see also an interesting domain patterns (see subplot b of Fig. 21). Narrow band domains having the direction of the ribbon axis are broken by the domains perpendicular to them. Such branched domain patterns [19] (i) have confirmed the presence of two magnetic phases and (ii) were observed practically at all annealed samples.

Surface magnetic behavior of SL samples is completely different. The hysteresis loops show low coercivity and low remanence-to-saturation ratio. This is typical for a magnetostrictive amorphous alloy with random internal stresses introduced during the rapid solidification. It is clearly seen that long-time annealing did not change much the shapes of loops. Moreover, in the domain patterns (Fig. 22) the existence of fingerprint domains and randomly oriented band domains coming from preparation process is still detected. This means that relaxation of local stresses was still not finished in SL ribbons.

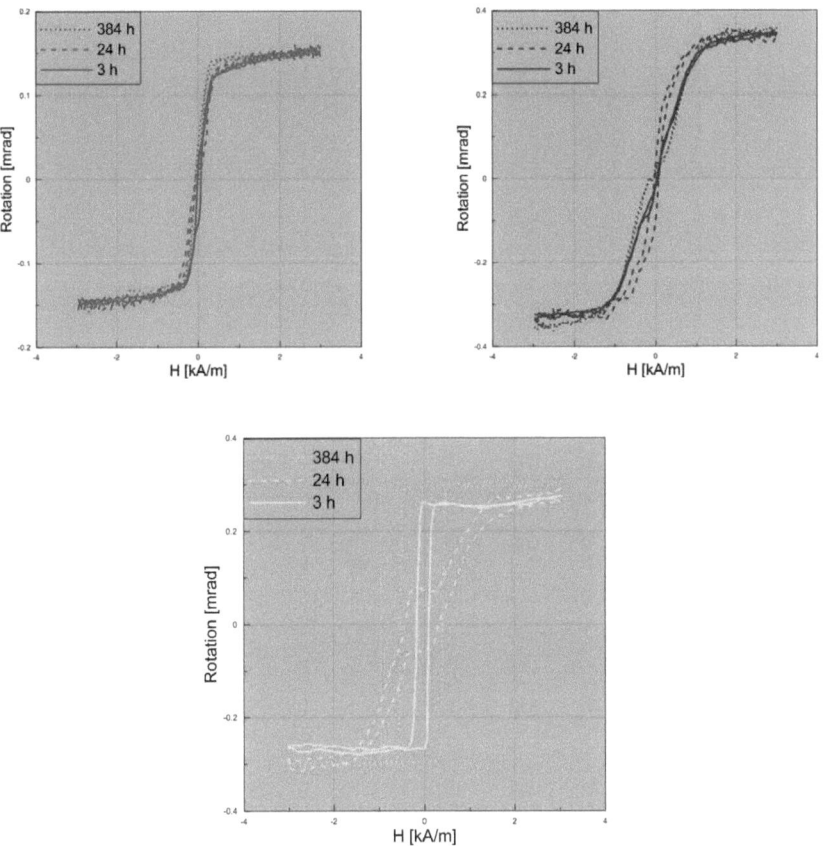

Figure 20: Magnetic surface hysteresis curves from the air side of the SL $Co_{59}Fe_{12}Cr_7Si_8B_{14}$ (upper left subplot), from the wheel of the SL $Co_{69}Fe_2Cr_7Si_8B_{14}$ (upper right subplot), and from the air side of the BL Co/Co (lower subplot) samples depicted as a function of annealing time.

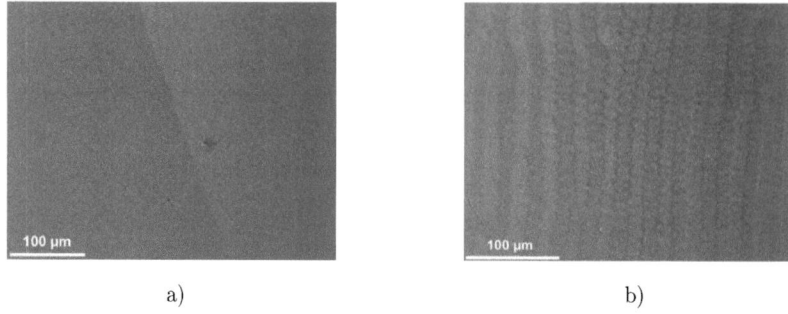

a)   b)

Figure 21: Observed domains at the all annealed Co/Co BL samples.

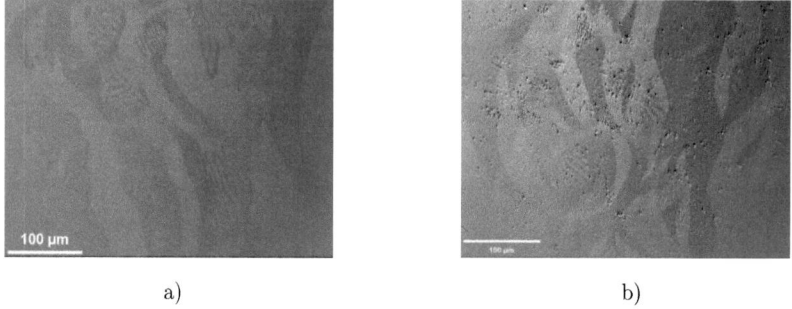

a)   b)

Figure 22: Magnetic domains at the air surface of the $Co_{59}Fe_{12}Cr_7Si_8B_{14}$ SL sample after 3 h of annealing (panel a) and 384 h of annealing (panel b).

4 Conclusions

My bachelor [11] and now diploma thesis are focused on detail investigation of new amorphous soft magnetic materials that can be mainly applied as stress or magnetic field sensors with tunable sensitivity. In bachelor work we have introduced PFC technology suitable for preparation of bi- or more layered ribbons with amorphous structure. Using this technology two different melt compositions, CoSiB and FeSiB, were used for production of the Co/Fe BL ribbon. Its surface, bulk, and interface physical and microstructural properties were analyzed and compared with the SL ribbons of corresponding compositions. We have identified also some negative aspects like coiling of the ribbon due to big differences in magnetostriction coefficients of layers, high fragility, still relatively high value of coercivity, etc. This resulted in the idea to change composition of both layers.

Present diploma work is devoted to the bilayered $Co_{69}Fe_2Cr_7Si_8B_{14}/$ $Co_{59}Fe_{12}Cr_7Si_8B_{14}$ (Co/Co) ribbon. Comparison of its physical and microstructural properties with properties of the Co/Fe samples and simultaneously again with SL ribbons of the consistent compositions have brought a more complex look at these new amorphous magnetic materials. Compared to the bachelor work we have extended our investigations about observations of the magnetic domain structure at ribbon cross-section and study of the structural and magnetic stability during long time temperature exploitation.

- Amorphous structure was confirmed by XRD on both sides of Co/Co ribbon and corresponding sides of SL samples. Changes in microstructure due to annealing process correlated well with cluster relaxation.

Long-time annealing at temperature 423 K evoked origin of surface crystallization.

- Thickness of interface at Co/Co sample is much thinner (< 1 μm) than that at Co/Fe sample (up to 6 μm). This lower thickness influences behaviour of microhardness, Young's modulus, and magnetic domain patterns at cross-section, and also the shape of bulk thermomagnetic curves.

- Surface magnetic properties of both BL samples are determined by the applied tensile/compressive stress induced due to unbending the ribbon on the planar sample holder. Induced magnetic anisotropy overcomes local anisotropies from the preparation process.

- Bulk magnetization and thermomagnetic curves indicate dominating role of air iron rich layers in the bulk magnetic properties of both BL samples.

- Annealing at temperature 423 K leads to the considerable softening of the magnetic properties with maintaining original magnetization saturation. On the other hand, fragility of the Co/Co sample stayed practically unchanged.

In the future we would like to devote our attention to other bi-layered ribbon type materials the composition of which should be reasonable from the viewpoint of magnetic properties but also from the mechanical (brittleness) ones. New possibilities offer combinations of the FeSiB/FeNbSiB and FeSiNbCuB/CoSiB bilayered systems. The experimental experiences obtained in present studies will be useful also in investigations of other type of materials prepared by PFC and/or other technologies.

References

[1] P. RIPKA, K. ZÁVĚTA K.H.J. BUSCHOV (Ed.). *Handbook of Magnetic Materials*, vol. 18North-Holland, Amsterdam, 2009, pp. 347–414 (Chapter 3)

[2] T. KLINGER, H. PFUTZNER, P. SCHONHUBER, K. HOFFMANN, N. BACHL. *Magnetostrictive amorphous sensor for biomedical monitoring*, IEEE Trans. Magn., 1992, vol. 28, pp. 2400–2402.

[3] E. KANIUSAS, L. MEHNEN, C. KRELL, H. PFITZNER. A magnetostrictive acceleration sensor for registration of chest wall displacements. *J. Magn. Magn. Mater.*, 2000, vol. 215–216, pp. 776–778.

[4] L. MEHNEN, H. PFUTZNER, E. KANIUSAS. Magnetostrictive amorphous bimetal sensors. *J. Magn. Magn. Mater.*, 2000, vol. 215–216, pp. 779–781.

[5] P. DUHAJ, P. SVEC, E. MAJKOVA, V. BOHAC, I. MATKO. *Mat. Sci. Eng. A*, 1991, vol. 133, pp. 662–666.

[6] L. KRAUS, V. HASLAR, K. ZAVETA, J. POKORNY, P. DUHAJ, C. POLAK. An amorphous magnetic bimetallic sensor material. *J. Appl. Phys.*, 1995, vol. 78, pp. 6157–6164.

[7] D. ATKINSON, P. DUHAJ. Magnetoelastic behaviour of amorphous bimetallic ribbons. *J. Magn. Magn. Mater.*, 1996, vol. 157–158, pp. 156–158.

[8] O. ZIVOTSKY, A. TITOV, Y. JIRASKOVA, J. BIRSIK, J. KALBACOVA, D. JANICKOVIc, P. SVEC. Full-scale magnetic, microstruc-

tural, and physical properties of bilayered CoSiB/FeSiB ribbons. *Journal of Alloys and Compounds*, 2013, vol. 581, pp. 685-692.

[9] A.MITRA,R.K. ROY,B. MAHATO, A.K. PANDA, G. VLASAK, D. JANICKOVIC, P. SVES. Development of FeSiB/CoSiB Bilayered Melt-spun Ribbon by Melt-spinning Technique. *Journal of Superconductivity and Novel Magnetism*, 2011, vol. 24, issue 1-2, pp. 611-615.

[10] D. IMAMURA, T. TODAKA, M. ENOKIZONO. Fe–Mn–Si/6.5 wt%Si–Fe Bilayer Ribbons Produced by Using the Melt-Spinning Technique. *IEEE Trans. Magn.*, 2011, vol. 47, pp. 3184–3187.

[11] A. TITOV. *Structure and properties of CoSiB/FeSiB alloys perspective for bilayered sensors.* VSB - Technical University of Ostrava, 2013.

[12] G. VLASAK. Direct measurement of magnetostriction of rapidly quenched thin ribbons. *Journal of Magnetism and Magnetic Materials*, 2000, vol. 215-216, pp. 479-481.

[13] P. VOJTANIK, J. KRAVCAK, R. VAGRA.Complex permeability after-effects in an annealed Co-based amorphous alloy. *Journal of Magnetism and Magnetic Materials*, 1996, vol. 157-158, pp. 175-176.

[14] A. TITOV, O. ZIVOTSKY, Y. JIRASKOVA, A. HENDRYCH, J. BURSIK P. SVEC. Influence of magnetostriction on cross-section magnetic properties in bilayered ribbons. *Magnetics, IEEE Transactions on*, 2014, vol. 50, issue 11, pp. 1-4.

[15] Y. JIRASKOVA, J. BURSIK, I. TUREK, M. HAPLA, A. TITOV, O. ZIVOTSKY. Phase and magnetic studies of the high-energy alloyed Ni–Fe. *Journal of Alloys and Compounds*, 2014, vol. 594, pp. 133-140.

[16] A. CHATURVEDI T. P. DHAKAL, S. WITANACHCHI, ANH-TUAN LE, MANH-HOUNG PHAN, H. SRIKANTH. Critical length and giant magnetoimpedance in $Co_{69}Fe_{4.5}Ni_{1.5}Si_{10}B_{15}$ amorphous ribbons. *Materials Science and Engineering: B*, 2010, vol. 172, issue 2, pp. 146-150.

[17] D. M. MINIC, V. A. BLAGOJEVIC, D. M. MINIC, T. ZAK. Influence of microstructural inhomogeneity of individual sides of Fe81Si4B13C2 amorphous alloy ribbon on thermally induced structural transformations. *Materials Chemistry and Physics*, 2011, vol. 130, issue 3, pp. 980-985.

[18] A.L. PATTERSON. The Scherrer Formula for X-Ray Particle Size Determination, *Phys. Rev.*, 1939, vol. 56, pp. 978-982.

[19] A. HUBERT, R. SCHÄFER. *Magnetic domains : the analysis of magnetic microstructures.* Springer, 1998.

List of author's publications

Refereed international papers:

1. O. ZIVOTSKY, A. TITOV, Y. JRASKOVA, J. BIRSIK, J. KALBA-COVA, D. JANICKOVIC, P. SVEC. Full-scale magnetic, microstructural, and physical properties of bilayered CoSiB/FeSiB ribbons. *Journal of Alloys and Compounds*, 2013, vol. 581, pp. 685-692.

2. Y. JIRASKOVA, J. BURSIK, I. TUREK, M. HAPLA, A. TITOV, O. ZIVOTSKY. Phase and magnetic studies of the high-energy alloyed Ni–Fe. *Journal of Alloys and Compounds*, 2014, vol. 594, pp. 133-140.

3. A. TITOV, O. ZIVOTSKY, Y. JIRASKOVA, A. HENDRYCH, J. BURSIK P. SVEC. Influence of magnetostriction on cross-section magnetic properties in bilayered ribbons. *IEEE Transactions on Magnetics*, 2014, vol. 50, issue 11, pp. 1-4.

4. Z. CHROMACKOVA, F. KOVANDA, D. LEGUT, A. TITOV, M. RITZ, D. FRIDRICHOVA, S. MICHALIK, P. KUSTROWSKI, K. JIRATOVA. Effect of precursor synthesis on catalytic activity of Co_3O_4 in N_2O decomposition. *Catalysis Today*, 2015 (in press).

Abstracts of international conferences:

5. A. TITOV, D. LEGUT, L. OBALOVA: An *ab initio* study of the phonon vibrations of Co_3O_4. Abstract of the conference NanoOstrava 2015, VŠB - Technical University of Ostrava, May 18-21, 2015.

6. A. TITOV, D. LEGUT, L. OBALOVA: Infra-red and Raman frequencies of Co_3O_4 – an *ab initio* study. Abstract of the 79th Annual Meeting of the DPG and DPG Spring Meeting, Berlin, 15-20 March 2015.

7. A. TITOV, O. ZIVOTSKY, Y. JIRASKOVA, A. HENDRYCH, J. BURSIK, P. SVEC. Influence of magnetostriction on cross-section magnetic properties in bilayered ribbons. Abstract of the IEEE International Magnetic Conference, Dresden, Germany, May 4-8, 2014.

Bachelor thesis:

8. A. TITOV. *Structure and properties of CoSiB/FeSiB alloys perspective for bilayered sensors.* VSB - Technical University of Ostrava, 2013.

Printed by Books on Demand GmbH, Norderstedt / Germany